Louiza Ouksel

Antioxidant effectof an α-aminophosphonate

Louiza Ouksel

Antioxidant effectof an α-aminophosphonate

ScienciaScripts

Cover image: www.ingimage.com

This book is a translation from the original published under ISBN 978-620-6-72916-7.

Publisher:
Sciencia Scripts
is a trademark of
Dodo Books Indian Ocean Ltd. and OmniScriptum S.R.L publishing group

120 High Road, East Finchley, London, N2 9ED, United Kingdom
Str. Armeneasca 28/1, office 1, Chisinau MD-2012, Republic of Moldova, Europe
Managing Directors: Ieva Konstantinova, Victoria Ursu
info@omniscriptum.com

Printed at: see last page
ISBN: 978-620-8-37033-6

Antioxidant effect of an α-aminophosphonate

Antioxidant / Free radical / Healthy skin

Signing sessions

- *I* dedicate this work to the memory of my mother, my father and my grandparents, may Allah the Almighty grant them peace and mercy.
- *To* my husband, who helped me a lot with his advice.
- *To* my daughters Imen, Amina and Maissa Nour el Houda.
- *To* my brother Mohammed and my sister Zahia.

FOREWORD

During metabolic processes, the body undergoes oxidation and reduction reactions which lead to the production of free radicals. The balance between the production and elimination of these radicals ensures the maintenance of the body's normal physiological functions. Excessive production, however, causes damage to biological molecules, leading to tissue damage and the onset of many diseases.

The body can be protected against oxidative damage from these molecules by antioxidants. For these reasons, the search for new molecules has become a major focus of research worldwide. In recent years, researchers have turned their attention to the synthesis of lipophilic compounds known as phosphonates. These compounds are characterized by broad biological and pharmacological activity, and thus the ability to reduce and/or treat diseases linked to oxidative stress.

The α-aminophosphonates in particular are considered an important class of phosphonates, with structures similar to those of amino acids. They are low-toxicity, biologically active compounds. This makes them ideal for biological and medicinal applications.

The aim of this book is to synthesize the new α-aminophosphonate acid derivative by conventional methods and microwave irradiation, identify it by various physico-chemical analysis methods and then demonstrate its antioxidant activity.

Chapter I: is devoted to the bibliographical study which presents general notions on α-aminophosphonates their synthesis methods and their applications in different fields.

Chapter II: describes the methods used to assess antioxidant activity, represented by a chemical method (DPPH test) and an electrochemical method (cyclic voltammetry).

Chapter III: we present and discuss the various experimental results obtained.

Table of contents

Signing sessions 3

FOREWORD 5

Chapter 1: α-Aminophosphonates 8

Chapter 2: Methods for assessing antioxidant activity. 21

Chapter 3: Results and discussion 25

References 39

Chapter 1: α-Aminophosphonates

1.1 Introduction

The history of phosphonates began in 1865 in Marburg, when Nikolai Menchutkin described the synthesis of bisphosphonates by a reaction between phosphorous acid and chloracetyl. Thanks to their physicochemical properties, phosphonates cover a wide range of applications in a variety of sectors.

1.2 α-aminophosphonates

A-aminophosphonate compounds are a specific family of widely distributed phosphonates, involved in many important biological processes. The presence of the nitrogen atom in a-aminophosphonates increases their chelating power towards metals and the stability of the complexes formed. Among this class of aminophosphonates, α-aminophosphonic acids are the most important compounds due to their amino acid-like structures, in which the carboxylic group -COOH is replaced by the phosphonic acid P(O)(OH)2.

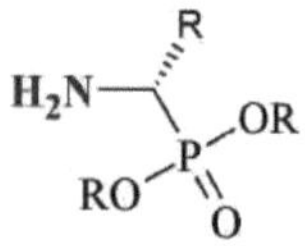

R
H_2N
OR
P
RO
O

Figure I.1: Structure of α-aminophosphonates.

Ouksel et al have synthesized three a-amino-phosphonate derivatives by theMoedritzer-irani method.

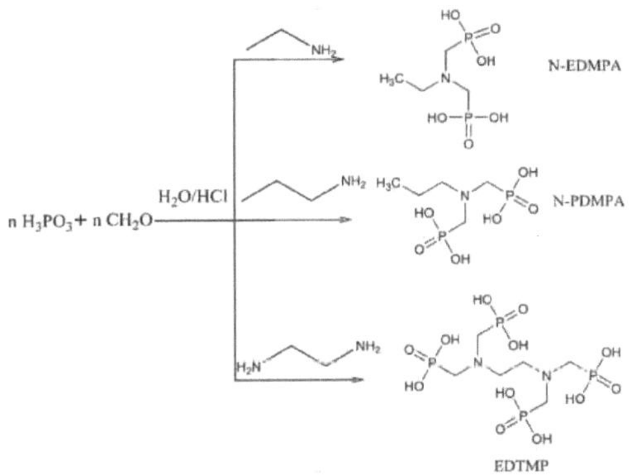

Figure I.2: Synthesis of acid *α-aminophosphonates*.

Tlidjane et al synthesized an α-aminophosphonic acid derivative by the reaction between amine and phosphorous acid and thiophenes carboxaldehyde, and subjected the mixture to microwave irradiation (100 w) for 3-8 min.

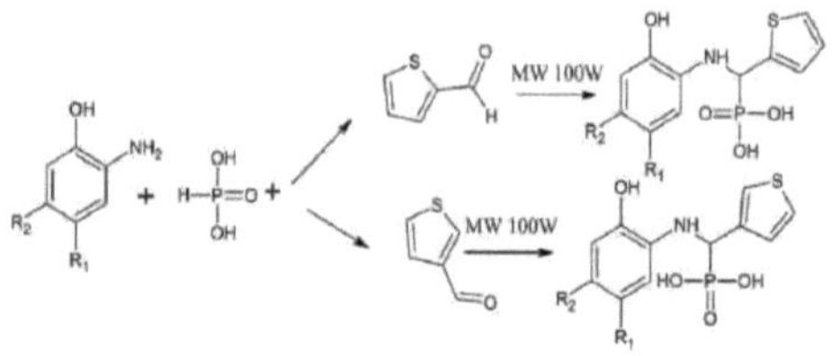

Figure I.3: Microwave synthesis of α-aminophosphonate acids.

Damiche et al, have synthesized a novel bioactive α-aminophosphonate by the three-component reaction between an electrophilic carbonyl compound (aldehyde) and a nucleophilic amine and diethyl phosphite. The synthesized products were prepared in the molar ratio 1/2/2 of amine, aldehyde and diethyl phosphite respectively.

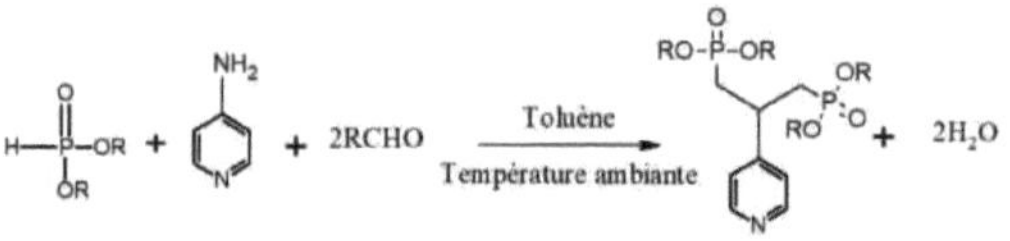

Figure I.4: Synthesis of α-aminophosphonate esters.

Varalakshmi et al have prepared a series of n-aminophosphonates using the Kabachnik-Fields reaction between three compounds 4-amino-2,6- dibromophenol, a series of heterocyclic/phenyl-substituted aldehydes and diethylphosphite, and catalyzed by $CeCl_3.7H_2O$ (5 mol%) under microwave irradiation.

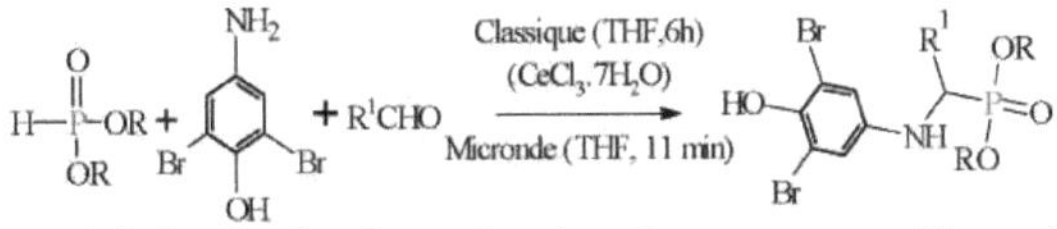

Figure I.4: Synthesis of α-aminophosphonate esters with catalyst.

I.3Applications of α-aminophosphonates

Phosphonates, and in particular α-aminophosphonates, have long been considered among the most widely used compounds in various fields of life. In the following, we present a few areas of application in biology and medicine.

I.3.1 Biology

α-Aminophosphonates are widely used in biology and as the most interesting compounds for human beings. They have diverse biological activities such as antibacterial, antifungal, antiviral, anti-inflammatory, anti-hypertensive, antitubercular, antioxidant and enzyme inhibitor.

α-Aminophosphonic acids have been used as intermediates in the synthesis of protease inhibitors and as "haptens" of catalytic antibodies.

Hellal et al have synthesized, characterized and evaluated in vitro the antibacterial and antifungal activity of a new series of three α-aminophosphonic acids.using four types of Gram-positive and four types of Gram-negative bacteria. This study shows that the compounds studied are very interesting antibacterial and antifungal agents.

α-Aminophosphonates are effective inhibitors of many enzymes, often those involved in amino acid metabolism.

Damiche and chafaa have studied the biological properties of a new series of bioactive molecules belonging to the α-aminophosphonate family, the results of which indicate that these molecules have significant antioxidant, anti-inflammatory and antibacterial properties and low toxicity.

I.3.2 In medicine

α-Aminophosphonates can be used as antioxidants, as shown by a study by Guna Subba Reddy et al, who prepared a series of α-aminophosphonate derivatives and evaluated their antioxidant activities. The results show that the molecules synthesized exhibit significant antioxidant activity.

Huang et al have synthesized a new series of α-aminophosphonates derived from dehydroabietic acid, and tested them in vitro against tumor cells. The results showed superior anticancer activity to that of 5-fluorouracil (the reference anticancer drug).

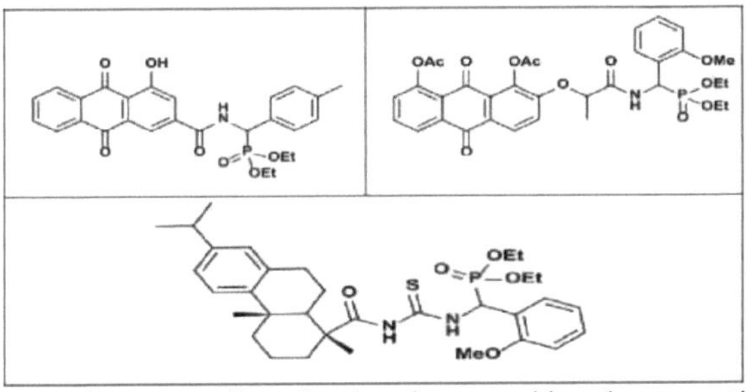

Figure I.5: Structures of a-aminophosphonates with anti-tumor activity.

I.4 Synthesis of α -aminophosphonates

1.4.1 Classic synthesis

2 mmol (0.432 mg) of amine (4.4'-TDA) are mixed in a bicol flask with 8 mmol (0.4 ml) of H3PO3, 10 ml of HCl and 20 ml of H_2 O. The mixture is stirred vigorously at 110°C for 1 h, then 16 mmol (0.3 ml) formaldehyde is added dropwise in a reflux set-up for 2 h, under an N_2 atmosphere. At the end of the reaction, the product is extracted with ethanol and recovered as a brick-red solid, washed several times with ethanol, filtered and dried.

1.4.2 Microwave synthesis

The use of microwaves (MO) to accelerate a chemical reaction is a valid alternative to other heating methods (reflux, oil or sand baths), the most widely used, and can be particularly effective where conventional processes are limited or ineffective.

Microwave activation results from two contributions, the first of purely thermal origin resulting from molecular agitation caused by the inversion of dipoles with alternating electric fields. The second, microwave irradiation heating, was developed in the field of organic synthesis following the work of Gedye and Giguerre in 1986.

The general protocol for microwave synthesis is to mix 1 mmol (0.216 mg) of the amine (4.4'- TDA) with 4 mmol (2 ml) of H3PO3 in the presence of 10 ml HCl and 10 ml d³⁄4O in an Erlenmeyer flask, subject the mixture to vigorous stirring and then to microwave irradiation at a power of 30W for two minutes, before adding 8 mmol (0.15 ml) of formaldehyde drop by drop. At the end of the reaction, the product is recovered with ethanol as a brick-red solid, washed several times and dried.

I.4.3 Evaporation

Rotary evaporation is a fast and efficient separation technique, using an apparatus based on simple vacuum distillation, which can rapidly remove large quantities of solvent, albeit only partially, often referred to as "Rotavap". The reaction mixture is placed in the upright flask, which is immersed in a water bath. The flask is tilted and rotated (100 RPM) to create a film of liquid and increase the surface area for solvent evaporation. The pressure inside the assembly is lowered by means of a pump, in our case a water tube, which increases the evaporation rate. After condensation in the cooler, the solvent is collected in the flask on the left.

I.5 Identification of the synthesized product

I.5.1Thin layer chromatography (TLC)

Thin-layer chromatography (TLC) is a physical separation method based primarily on the chemical affinity of two phases, one stationary and the other mobile: the stationary phase is a solid fixed to an aluminum, plastic or glass plate. The mobile phase is a solvent or solvent mixture, which moves along the stationary phase. Once the solute has been deposited on the stationary phase, the substances migrate essentially from the base of the plate upwards by capillary action, at a speed which depends on the electrostatic forces holding the component to the stationary phase, and on its solubility in the mobile phase, characterized by a frontal ratio (Rf).

To follow the evolution of our reaction, we used TLC plates, whose size varied according to the number of products to be tested. Product spots are deposited on these plates. The eluent used for this reaction is methanol and dichloromethane (4/1, v/v) for clear migration. The plates were developed using a UV lamp set to [254, 365 nm].

I.5.2 UV-vis spectrophotometry

Ultraviolet and visible absorption spectroscopy (UV-vis) is a quantitative and qualitative analysis method. It is based on the absorption of light energy by a substance in the 200-800 nm range. When a molecule absorbs a portion of the energy of electromagnetic radiation, it is automatically accompanied by an electronic transition from a fundamental level to a higher energy level. These electronic transitions, occurring on a molecular scale, involve valence electrons. Energy absorptions are expressed by Beer-Lambert's law. This law is used to measure absorbance A, or transmission T %.

$$T(\%) = \frac{I}{I_0}$$

$$A = \varepsilon . l . c$$

Where:

I0: Intensity of incident light.

I: Intensity of transmitted light.

C: Solution concentration in mol/l.

T: Transmission percentage.

A: Absorbance.

In our work, UV-Vis spectra of the synthesized product 4,4'- TDMBP and its starting product (4,4'-TDA) were recorded in DMSO using 1 cm-wide quartz cells filled with a solution of either the synthesized product or the starting product in DMSO, at room temperature in the range from 200 nm to 900 nm.

I.5.3 Infrared (IR) spectrometer

An IR spectrometer produces a basic document called an infrared spectrum. IR spectra are recorded by plotting the inverse of the wavelength (wavenumber), expressed in cm, on the x-axis^{-1} . The ordinate shows the transmittance or percentage (T%) of each radiation. Each band is characterized by its absorption maximum value; its relative intensity is also specified (F: strong, m: medium, f: weak).

FT-IR spectra of the synthesized compound 4,4'- TDMBP and its starting product 4,4'- TDA were recorded in solid phase on a JASCO FT/IR-4200 spectrometer. Spectra were recorded in the 4000-500 cm range .$^{-1}$

I.5.3 Solubility tests

Solubility is a quantitative method and one of the characteristic physical properties of chemicals, signifying the ability of a solute to dissolve in a solvent, to form a homogeneous solution. Solubility tests were carried out in the laboratory, using a range of solvents (water Methanol, Ethanol, DMSO, Chloroform, Ethyl acetate, Dichloromethane) and a quantity of synthesized 4,4'-TDMBP or starting material 4,4'-TDA with stirring.

Table III.3: Solubility test results for 4.4'-.

TDA and A4.4'-TDMBP.

Solvant	4,4'-TDA	A4,4'-TDMBP
Eau	-	-
Ethanol	+	-
Méthanol	+	-
(DMSO)	+	+
Chloroforme	+	-
Dichlorométhane	+	-
Acétate d'éthyle	+	-

(-: insoluble, +: soluble)

The solubility test we carried out showed that our product was not soluble in any of the solvents tested, with the exception of DMSO.

I.5.5 Melting point

The melting point is the temperature at which a body changes from a solid to a liquid state. It is a characteristic physical property of matter that provides essential information on the composition and purity of materials. The melting point of the synthesis product (4.4'-TDMBP) was determined using a Fusiometer (BUCHI) Melting point B-540, a measuring device with a temperature gradient ranging from 20 to 900°C. The product is placed directly into capillary tubes, without the need for any accessories. When heated from a given point, a change in texture appears in the sample.

Table III.4: Melting points of synthesized products and their starting materials.

Produit	Point de fusion (°C)
4,4'-TDA	105-107
A4,4'-TDMBP	172

Chapter 2: Methods for assessing antioxidant activity.

II.1 Chemical method (DPPH)

Antioxidant activity is determined using the stable radical 2,2-diphenyl-1-picrylhydrazyl (DPPH·). The DPPH radical is purple in its oxidized form, if an antioxidant is added it turns yellow its reduced form is called 2,2-diphenyl-1-picrylhydrazine (DPPH-H).

DPPH- + AH → DPPH-H + A-

Or (AH) is a compound capable of yielding an II to the DPPH· radical.

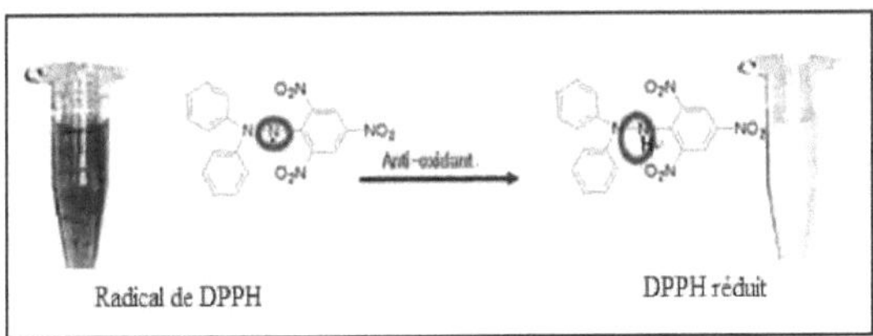

Figure II. 1: Reduction of the DPPH radical.

DPPH· is widely used to test the free radical scavenging capacity of the compound under test, with its absorption maximum at around $\lambda = 517$ nm in methanol and ethanol.

The hydrogen transfer reaction from antioxidant to DPPH* is monitored by visible spectroscopy, recording the decay of the DPPH* absorption band at $\lambda = 517$ nm (decrease in violet coloration). This test can be used to determine several parameters enabling antioxidants to be compared with each other. The IC50, a parameter that defines the effective concentration of a compound that causes the loss of 50% of DPPH* radical activity. The lower the IC50 value, the more powerful the compound is considered to be as an antioxidant (Molyneux, 2004). To measure this activity, we used the method described by Que (Que et al. 2006). A methanolic solution of 0.04 mg/ml DPPH is mixed with different concentrations of the synthesized molecule (1mg/ml) diluted with methanol/DMSO (50-800 µg/ml), we add 150 µl of each dilution of this product(4,4'-TDMBP) into a test tube, then we add 750 µl of methanolic DPPH solution, shake well with a vortex then let incubate 30 min protected from light at room temperature. We then read the absorbance at 517 nm against a blank containing methanol/DMSO. We repeat the same operations, replacing the synthesized product (4,4'-TDMBP) with ascorbic acid (positive control). The negative control consists of 0.5 ml DPPH methanolic solution and 0.5 ml methanol. All operations are performed in duplicate. Evaluation of antioxidant activity using the DPPH method is expressed as a percentage of inhibition according to the following relationship:

$$\%\ d'inhibition = \frac{A_{DPHH} - A_{Echantion}}{A_{DPPH}} \times 100$$

Where:

A_{DPPH}: absorbance of the control reaction alone.

A_{Eample}: absorbance in the presence of the test product.

II.1 Electrochemical method (VC)

Cyclic voltammetry (CV) is one of the most effective electrochemical methods for characterizing charge transfer reactions. It involves imposing a linear ramp in potential with a positive or negative sweep rate, and recording the current intensity. The set-up generally used consists of three electrodes (a glassy carbon working electrode, an Ag/AgCl reference electrode, a platinum plate auxiliary electrode, an electrochemical cell, and an electronic circuit, called a potentiostat, for modifying the potential and recording the current.

Before starting to plot voltammograms, we prepare the supporting electrolyte, which is tetrabutylammonium tetrafluoro borate (TBu4NBF4) solubilized in DMSO at a concentration of 0.1 M. Measurement solutions are prepared from a stock solution prepared by dissolving 4.4'-TDMBP test material (0.06 g) in 10 ml DMSO. The working electrode is stripped by polishing with a paper containing a diamond paste, then rinsed with distilled water and DMSO before each manipulation. We saturate the study solution with O2 bubbling for 5 min before each manipulation. Measurements were performed on a 10 ml sample volume. Cyclic voltammograms are recorded after 10 min of o_2 bubbling. Repeat the same operations, replacing the synthesized product with ascorbic acid. Before starting the study of our product, we specified the electroactivity range of the supporting electrolyte [-1200, - 400 mv] with a scan speed of 50 mv/s.

Figure II. 2: Experimental set-up used for voltammetric measurements.

The percentage of radical scavenging activity with respect to superoxide radicals of compounds was calculated using the following equation:

$$I\% = \frac{I^0_{pa} - I_{pa}}{I^0_{pa}} X100$$

Where:
$I^0{}_{pa}$: Anodic peak current for O2 oxidation without test compound.

I_{pa}: anodic peak current for O2 oxidation with the compound under test

Chapter 3: Results and discussions

III.1Synthesis of A4,4'-TDMBP

The main direct route to α-aminophosphonic acid is a one-pot reaction involving three components, two of which are nucleophilic and the third electrophilic, known as the Irani-Moedritzer reaction. It involves a carbonyl compound (aldehyde), an amine and a phosphorous acid in an acidic medium (HCl), according to the reaction illustrated in Fig. III. 1.

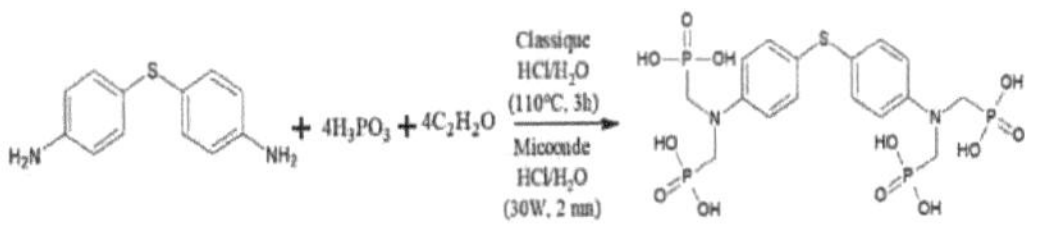

Figure III.1: Schematic diagram of the A4,4'- TDMBP synthesis reaction.

III.2 Reagents used

The physicochemical characteristics of the reagents used in the synthesis of A4,4'-TDMBP are shown in Table III.1.

Table III.1: Physicochemical properties of the reagents used.

Réactifs	Formule brute	M (g/mol)	Densité (g/cm^3)	Pureté %
Eau	H_2O	18.01	1	100
Acide phosphoreux	H_3PO_3	82.00	1.65	97
Acide chlorhydrique	HCl	36.45	1.19	37
Formaldéhyde	CH_2O	30.02	0.81	37
4,4'-Thiodianiline	$C_{12}H_{12}N_2S$	216.30	1.26	98

III.3 Characterization of A4,4'-TDMBP

III.3.1 Thin layer chromatography (TLC)

We followed the progress of the chemical reaction on a silica gel-coated TLC plate developed with an eluent mixture of methanol and dichloromethane in the ratio (4/1, v/v). Analysis of the plate under a UV lamp enables the end of the reaction to be estimated and the purity of the product to be tested. In our case, the appearance of a single spot for the synthesized product, which is different from that of the starting product, confirms the end of the reaction and the obtaining of a new product. The TLC plate shows the disappearance of the starting product spot (amine), which is apolar, whereas the synthesized product spot is not displaced, which explains why the compound is polar.

The frontal ratio (Rf) or retention factor of a compound is the ratio of the deposition line-compound distance to the deposition line-solvent front distance it is between 0 and 1 and is characteristic of the compound, the plate material and the elution system (Boulekras, 2014). From the TLC plate, we determined the front ratio (Rf) by the following equation:

Table III.2: Frontal ratios of starting product and synthesized compound.

The products	Rf
4,4'-TDA	0.66
A4,4'-TDMBP (MO)	0.00
A4,4'-TDMBP (Classic)	0.00

We note that the Rf of the starting product is totally different from that of the synthesized compound, indicating a successful synthesis. Next, we'll carry out a more detailed analysis (e.g. NMR, MS...) to ensure that the compound obtained is indeed the expected product.

III.2.1 Solubility tests

After synthesis, qualitative solubility tests were carried out on the synthesized product and the starting material in the available solvents. The results obtained are summarized in Table III.3.

Table III.3: Solubility test results for 4.4'-TDA and A4.4'-TDMBP.

Solvent	4,4'-TDA	A4,4'-TDMBP
Water	-	-
Ethanol	+	-
Methanol	+	-
DMSO	+	+
Chloroform	+	-
Dichloromethane	+	+
Ethyl acetate	+	+

(-: insoluble, +: soluble)

The solubility test we carried out showed that our product was soluble only in DMSO, dichloromethane and ethyl acetate.

III.3.3 Melting point

The melting point is a characteristic property of solid substances. It is the temperature at which the solid state changes to the liquid state. We used this analysis to determine the melting point of the new compound we obtained.

The melting point results for the synthesized compound and the starting material are shown in Table III 4, which indicates that the melting point of the synthesized compound is different from that of the starting material.

Table III.4: Melting points of synthesized products and their starting materials.

Product	Melting point (°C)
4,4'-TDA	105-107
A4.4 -TDMBP'	172

III.2.2 Characterization by UV-vis spectrophotometry.

UV-vis spectra of the synthesized product and the starting material are shown in Figures III.3 and III.4.

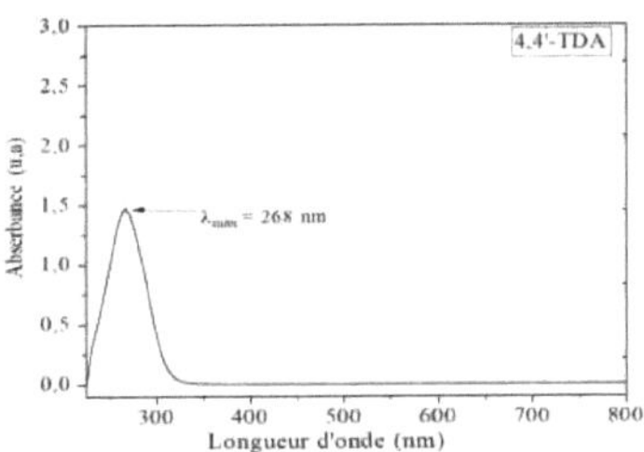

Figure III.3: UV-vis absorption spectrum of 4,4'-TDA in DMSO.

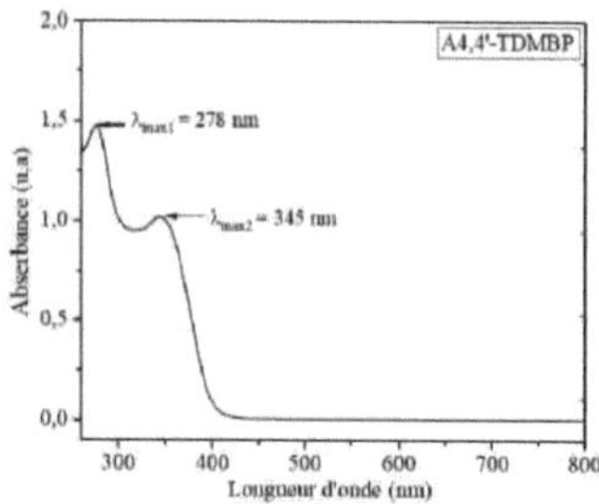

Figure III.4: UV-vis absorption spectrum of A4, 4'- TDMBP in DMSO.

From these spectra, we can see a clear difference between the spectrum of the synthesized product and the spectrum of the starting product represented by the creation of other bands on the spectrum. The UV-vis spectrum recorded for the starting product 4,4'-TDA is characterized by an absorption band located at λ max = 268 nm, which is due to aromatic systems or electronic excitation in this region. In the visible range, this compound is transparent and shows no absorption.

The UV-vis spectrum recorded for A4,4'-TDMBP is characterized by two absorption bands: a strong band at λ_{max1} = 278 nm, which is due to aromatic systems or π →π* electronic excitation in this region, and a medium band at λ_{max2} = 345 nm that corresponds to the n→π* electronic transition. For our molecule, this type of transition is linked to the presence of heteroatoms (O, N and P) carrying free electron doublets.

III.2.3 Characterization by infrared spectroscopy

The IR spectra of the starting product and the synthesized molecule are shown in Figures III.5 and 6.

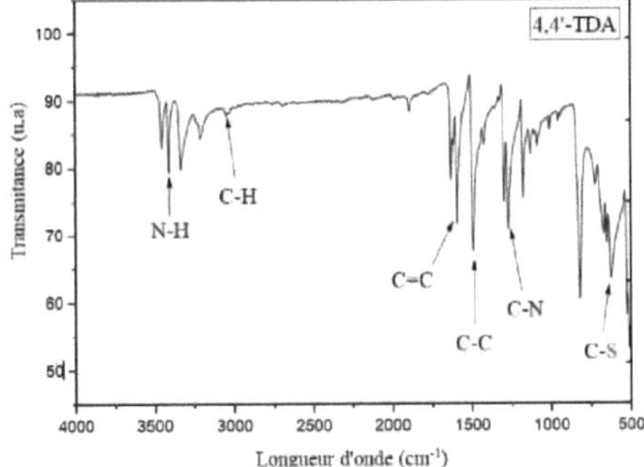

Figure III.5: Characteristic IR spectra of 4,4'-TDA.

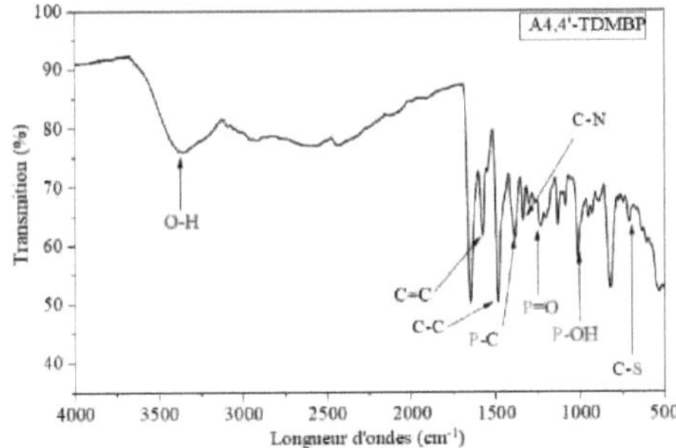

Figure III.6: Characteristic IR spectra of A4,4'- TDMBP

The IR spectra of the compound A4,4'-TDMBP and its starting product 4,4'-TDA show the existence of peaks characteristic of the functional groups, which are illustrated in Table III.5. The IR spectrum of 4,4'-TDA (figure III.5) shows an absorption band at 3404 cm-1 due to the elongation vibration of the primary amine group (NH_2). Absorption bands characteristic of functions present in compound 4,4'-TDA such as C-S, C=C, C-N, C-C, and C-H and their assignments are grouped together in Table III.5. The IR spectrum of A4,4'- TDMBP (figure III.6) obtained shows a broad absorption band at 3359 cm^{-1} due to the elongation vibration of the hydroxyl group (OH) and the appearance of band 1014 cm^{-} 1 attributed to the (P-OH) group, and the band corresponding to the elongation mode 1234 cm^{-1} attributed to the (P=O) which confirms the obtaining of α- aminophosphonic acid. In conclusion, the total disappearance of the NH2 primary amine function on the spectrum of the synthesized molecule and the appearance of the P=O, P-OH, P-C groups undoubtedly confirms the successful synthesis of α- aminophosphonic acid.

Table III.5: Main infrared frequencies (cm^{-1}) of synthesized product and starting material.

4,4'-TDA	A4,4'-TDMBP	Domaine/littérature	Assignements
/	3359	(3700-3125)	O-H
/	1400	(1445-1405)	P-C
1266	1333	(1340-1250)	C-N
/	1234	(1240-1150)	P=O
/	1014	(1040-910)	P-OH
614	700	(700-570)	C-S
1605	1582	(1600-1500)	C=C
3404	/	(3530-3400)	N-H
3037	3330	(3100-3000)	C-H
1490	1476	(1600-1450)	C-C

III.4 Assessment of antioxidant activity

III.4.1 Chemical method (DPPH)

Free radical scavenging activity is assessed by the DPPH method, which is frequently used for its simplicity and rapidity in estimating the free radical scavenging activity of a compound. This method is based on the reduction of an alcoholic solution of DPPH in the presence of an antioxidant that donates a hydrogen or an electron, the non-radical form DPPH-H is formed.

The antioxidant activity of the product obtained with respect to the DPPH radical was evaluated spectrophotometrically, by monitoring the reduction of this radical. To do this, we used a spectrophotometer to measure the absorbances of the samples at $\lambda = 517$ nm. This reduction capacity is determined by a decrease in absorbance induced by anti-radical substances.

The results obtained enable us to plot the variation of the percentage of inhibition as a function of the concentration of molecule synthesized and ascorbic acid $I\ \% = f(C)$ (figures III.7 and III.8).

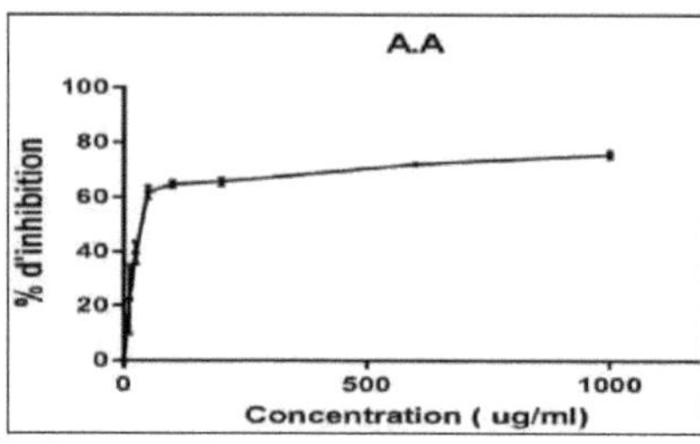

Figure III. 7: Radical scavenging activity of **A-A** as a function of concentration.

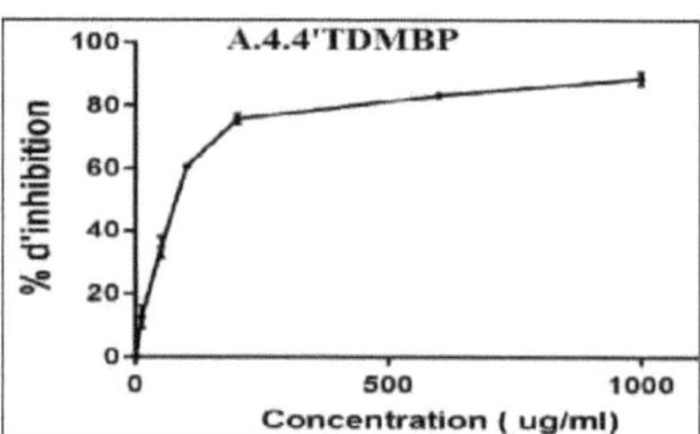

Figure III.8: Radical scavenging activity of A4,4'- TDMBP as a function of concentration.

These figures show an increase in the percentage inhibition of the synthesized product and ascorbic acid, with increasing concentration. To better characterize the free radical scavenging power of ascorbic acid, the IC50 parameter was introduced. This value was determined graphically from the curves representing the variation in percentage inhibition as a function of the concentration of the compound studied, and compared with that of the standard antioxidant (A.A). The IC50s of the product studied and that of A.A were determined and are presented in Table III.6.

Table III.6: Comparison of IC50s for the synthesized product and ascorbic acid.

Produit	A-A	A4,4'-TDMBP
IC50 (µg/ml)	25,08 ± 3.88	64.82 ± 4,64

For a better comparison between the antioxidant activity of the synthesized molecule and the standard, we have presented the CI 50 values in the form of a histogram (Figure III.9). Generally speaking, the lower the CI 50 value, the higher the antioxidant activity of a compound.

For the same concentration, the IC50 of our product is close to that of ascorbic acid. The antioxidant activity of A4,4'-TDMBP can be explained by the presence of both phosphonate groups (O=P (OCH2CH3) and the (C-N) group in its molecular structure.

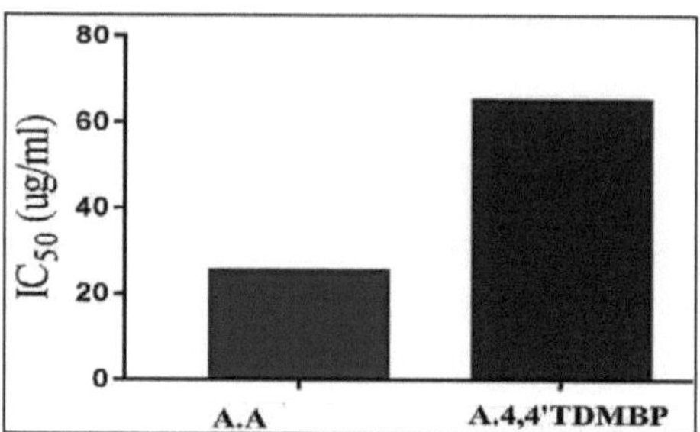

Figure III.9: IC50 of the DPPH free radical scavenging activity of A4,4'-TDMBP and A-A.

III.4.2 Electrochemical method

Cyclic voltammetry is used to determine a radical's antioxidant activity. This method is based on the electrochemical generation of $O^{\bullet -}$ by the reduction of dissolved oxygen in the solvent DMSO. Superoxide $O^{\bullet}$ is formed by the reduction of one electron of dioxide O_2 , according to the following reaction:

$$O_2 + 1e^- \rightarrow O_2^{\bullet -}$$

III.4.1.1 Cyclic voltammetry of oxygen alone

Oxygen cyclic voltamograms are recorded in an electrochemical cell containing 20 µl of electrolyte solution (Bu4NBF4 (0.1M)/DMSO), saturated with O_2. The solubility of oxygen in DMSO is 2.1 mM. Voltamograms are recorded at a sweep rate of 50 mv/s over a range of -1000 to -300 mV/SCE at room temperature. Figure III.10 shows the cyclic voltammogram of oxygen in the Bu4NBF4/DMSO medium with a speed of 50 mV/s, in the absence of the synthesized compound.The cyclic voltammogram of the superoxide anion radical shows a reversible process with oxidation and reduction peaks with $\Delta Ep = 70$ mV, which is in good agreement with work cited in the literature (Vasudevan et al., 1995).

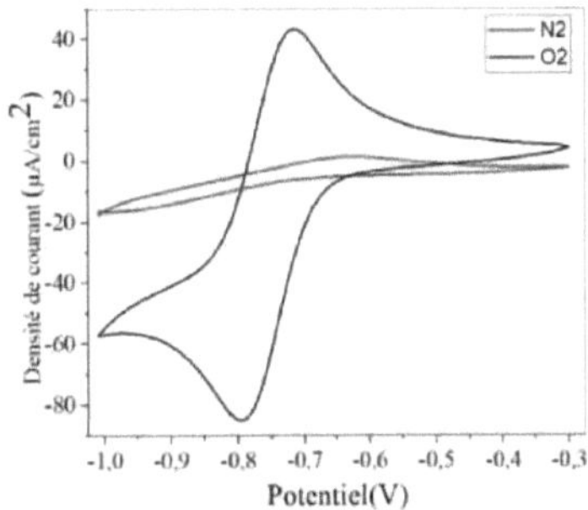

Figure III.10: Oxygen voltammetry in Bu4NBF4/DMSO medium at 50 mV/s.

III.4.1.2 Oxygen cyclic voltammetry in the presence of test compounds

The compound tested must be inactive in the potential range of the $O_2/O^{\bullet}$ couple⁻ to avoid interference or complex interpretation of the voltamogram. For this reason, before studying the radical trapping of our compound with respect to the superoxide radical, their cyclic voltammetry was carried out, and the latter was inactive in this potential range. Changes in the shape of the cyclic voltamograms of A-A and A4,4'-TDMBP are comparable. The cyclic voltamogram of the superoxide anion radical in the presence of A- A and A4,4'- TDMBP showed a decrease proportional to the anodic current, while the cathodic current intensity did not change significantly (Figure III.11).

The decrease in the anodic current of the superoxide anion (Figure III .11) suggests that the products tested are reacting irreversibly with this radical, while no change in the cathodic peak is observed, indicating the absence of interactions between the compounds tested and molecular oxygen. The superoxide anion radical formed electrochemically, reacts with the antioxidant (ROH) and forms an electroinactive species in the form of a radical and a conjugate base of hydrogen peroxide, on the other hand, the antioxidant base taking a proton to form a relatively stable product, RO·, is assumed to be of low toxicity due to the high probability of its dimerization.

$$O_2^{\bullet -} + ROH \rightarrow (RO - HO_2^{\bullet}) \rightarrow HO_2^{-} + RO^{\bullet}$$

$$HO_2^{\bullet} + ROH \rightarrow H_2O_2 + RO^{\bullet}$$

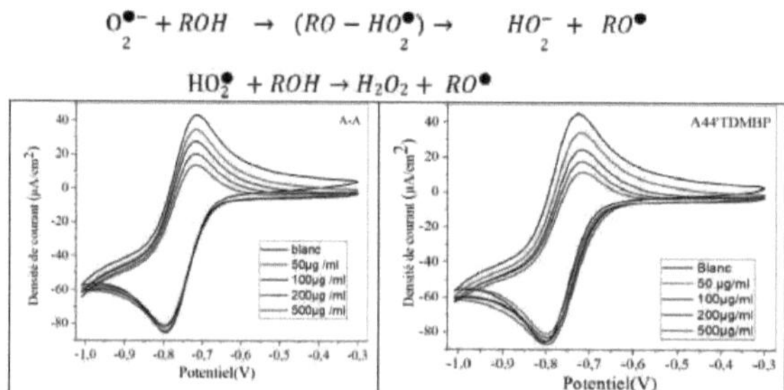

Figure III. 11: Evolution of cyclic voltamograms $O^{\bullet -}$ in the presence of different concentrations of A- A and A4,4'-TDMBP

References

- Abramov (1952). Reaction of dialkyl phosphites with aldehydes and ketones. A new method of preparation of esters of hydroxyalkanephosphonic acids. Zh Obshch Khim, 22:647 - 652.
- ChristopherR, Strauss, Rajender S, Varma.(2006) Microwave in Green Sustainable Chemistry, Top Curr Chem 266 :199-231.
- DamicheR. (2017). doctoral dissertation in pharmaceutical process engineering. Université Ferhat Abbas sétif-1.
- Mehri M, Chafai N, Ouksel L ,Benbouguerra K, Hellal A , Chafaa S. (2018). Journal of Molecular Structure 1171: 179-189.
- L. Ouksel, R. Bourzami, S. Chafaa, N. Chafai, Journal of Molecular Structure, 1222 (2020) 128813-128825.
- L. Ouksel, R. Bourzami, N. Hamdouni, A. Boudjada, Journal of Molecular Structure, 1229 (2021) 129792129803.
- R. Bourzami, L. Ouksel, N. Chafai, Journal of Molecular Structure 1195 (2019) 839-849.

Printed by Books on Demand GmbH, Norderstedt / Germany